CRYONICS

Navigating the Future of Life Sciences through Timeless Exploration

Anya Ellis

0

CRYONICS

Navigating the Future of Life Sciences through Timeless Exploration

Anya Ellis

Introduction

Cryonics, the process of preserving persons at extremely low temperatures after legal death with the possibility of future resuscitation, has moved beyond science fiction and into the realm of genuine scientific investigation. While the idea may appear daring, it is based on well-established concepts of cryopreservation, a technique that has been shown to be effective in maintaining cells and tissues at extremely low temperatures.

Death, which was always thought to be a definitive endpoint, is actually a more complicated phenomenon at the cellular level. Cellular death, or the collapse of individual cells, does not happen instantly. This window of time, during which some biological processes may still be active, provides a potential chance for intervention - the foundation of cryonics. Proponents think that by rapidly chilling the body to cryogenic temperatures (about -196°C), they can effectively pause these biological processes, perhaps allowing future medicine to intervene and restore life.

The cryonics method is a lengthy and complex

undertaking. It usually entails prompt medical intervention following legal death in order to minimize cellular damage. Blood is replaced with cryoprotectants, specialized chemicals that prevent ice formation within cells, which might cause irreversible harm. The body is then gradually chilled and eventually vitrified, resulting in a glass-like state. Finally, it is placed in a specialized cryostat filled with liquid nitrogen, where it awaits possible resuscitation in the future.

Chapter One

Historical Development

Humans' concern with sustaining life after death dates back to prehistoric cultures. The complex mummification processes used by the Egyptians, for example, are considerably different from cryonics but share a same thread: the desire to transcend the confines of mortality.

However, the scientific foundation for cryonics was established considerably later, in the mid-twentieth century. Advances in cryobiology, the study of life at low temperatures, paved the way for Robert Ettinger's innovative ideas. Inspired by the successful cryopreservation of cells and tissues, Ettinger postulated in his seminal article "The Prospect of Immortality" (1962) that this technology could be applied to humans, establishing the framework for current cryonics.

The 1960s saw the first tentative advances toward human cryopreservation. In 1967,

James Bedford became the first person to undergo the surgery, albeit under less-than-ideal conditions due to logistical restrictions. These early efforts, however, were hampered by the infancy of cryopreservation technology and a dearth of long-term data on its effects on the human body, creating serious ethical problems.

The following decades witnessed the formation of dedicated cryonics organizations, as well as advances in cryoprotectants, which are specialized compounds meant to reduce cellular harm during freezing. However, the field did suffer its share of problems. Financial limits, continued scientific uncertainty, and complex legal difficulties have limited cryonics' expansion and acceptance in the mainstream.

Chapter 2

Immortal Exploration

For millennia, humans have sought to transcend mortality, to avoid the inevitable. The concept of immortality, or the endless continuation of individual existence, crosses cultural and theological barriers, permeating mythology, literature, and philosophical discourse. However, when we go into this complicated concept, it is critical to distinguish between the realm of imagination and the possibilities, however distant, that science may provide.

Immortality is not a unified concept. It can take many different forms, each with its own set of ramifications. Certain jellyfish and single-celled animals exhibit Biological immortality, or the ability to postpone death indefinitely. However, for humans, this type of immortality is still considered theoretical biology.

Religious and spiritual traditions frequently propose the persistence of consciousness or soul after physical death, providing comfort and hope for a hereafter. While these beliefs have significant personal meaning for many people, they are outside the scope of scientific research.

Technological advancements in domains such as artificial intelligence and cryonics may provide views of potential pathways to a future in which the human experience can be extended, possibly by transferring consciousness to digital platforms or conserving bodies for future revival. However, these possibilities are riddled with scientific uncertainties, ethical quandaries, and philosophical issues.

Our interest with immortality arises from a variety of sources. The innate fear of the unknown, the need to see the future unfold, and the desire to leave a lasting legacy all add to the attraction of defying death. However, it is critical to recognize that mortality is an inherent part of the human condition. It influences our values, relationships, and experience of time.

Philosophical and Ethical Considerations

At its core, low-temperature biostasis calls into question traditional concepts of death and raises significant issues about the essence of existence. Is a person actually deceased if they are preserved after a legal death? Does the possibility of future revival change their existential status? These intricate challenges raise sensitive issues about the right to die and people' ability to choose their own fate, even outside conventional boundaries.

The high expense of low-temperature biostasis raises problems regarding access and fairness. Is it ethical to invest money in an unproven technology with uncertain advantages, thereby increasing existing gaps in healthcare systems? This subject involves a careful examination of resource allocation and the ethical implications of prioritizing this technology above urgent healthcare needs.

Individuals who have been preserved today may awaken to a completely different society, one with unexpected advancements and sociological shifts. Could they adjust to this new reality? Furthermore, is it ethical to make such a life-altering decision for someone who cannot participate in the discussion or consent to perhaps being revived in an unknown future?

The pursuit of low-temperature biostasis can be interpreted as a rejection of mortality, which is a basic feature of human existence. However, some say that accepting our mortality affects our beliefs, drives us to make the most of our limited time, and promotes a greater respect for life. Is the pursuit of unlimited life extension reducing the importance of our time on Earth and the events that shape us?

As the science of low-temperature biostasis advances, continual debate and critical reflection are essential. Open communication among scientists, ethicists, philosophers, and the general public is required to traverse the ethical complexity and ensure that this technology is developed and implemented responsibly.

Cultural Perspectives on Life Extension

The human urge to lengthen life and transcend the inevitable has left an indelible impression on societies all across the world. From ancient mythology to modern scientific discoveries, the concept of life extension evokes intrigue, hope, and, at times, serious societal and ethical problems. Understanding the many cultural viewpoints on this complicated problem is critical for developing educated discourse and managing any ethical issues that may arise.

Across cultures, the concept of life extension takes numerous forms. Eastern philosophies such as Taoism and Hinduism frequently emphasize spiritual longevity, which focuses on building inner calm and reaching a condition beyond the constraints of the physical body. Indigenous societies around the world may believe in **reincarnation or cyclical rebirth, which provides a unique perspective on the concept of individual lifespan.

Many Abrahamic religions believe that life is a sacred gift from a higher power, and death is a natural transition to an afterlife. The concept of extending life through technology may consequently be interpreted as tampering with the divine design. Other religious and moral frameworks may emphasize the dignity of a natural lifespan, raising ethical concerns about artificially extending life beyond its perceived natural course.

As scientific advances in medicine and bioengineering continue, the idea of extending human lifespan becomes more realistic. However, cultural viewpoints on life extension involve a wide range of considerations. Some may see it as an opportunity to contribute to society by living longer, healthier lives, while others may be concerned about resource allocation, societal consequences, and the potential upheaval of established cultural norms and traditions.

Chapter 3

The Future of Life Sciences

Researchers are actively developing innovative cryoprotectant formulations using a variety of methodologies. Nanotechnologies, for example, show promise for developing tailored delivery systems that can safeguard critical cellular components. Furthermore, study into the natural antifreeze qualities seen in certain extremophiles (organisms that flourish in harsh conditions) may provide useful insights for building more efficient cryoprotectants.

Another problem is the freezing procedure itself. While freezing the body may appear to be a simple solution, it is not optimal for cryonics due to the risk of cellular damage induced by ice formation. vitrification is a potential remedy. By progressively eliminating water and replacing it with cryoprotectants, the body is turned into a glass-like state, preventing the creation of ice crystals entirely.

Biotechnology is critical in advancing this endeavor: Advanced microscopy techniques such as cryo-electron microscopy provide an unprecedented insight into the cellular world at high resolution. This enables researchers to monitor the effects of vitrification on various cell types and improve the technique to reduce cellular harm. Furthermore, research into biocompatible artificial polymers with certain features may help to develop novel vitrification treatments.

While the prospective applications of biotechnology in cryonics provide a look into the future, it is important to recognize that the field is still in its early phases. Significant scientific and technological improvements are required before it can be considered a feasible option. Navigating this complicated landscape will require open collaboration among researchers in cryonics, biotechnology, and other related domains, guided by a commitment to scientific rigor and ethical considerations.

Nanotechnology and Its Impact

The field of cryonics, which seeks to preserve persons at ultra-low temperatures for possible future resuscitation, presents substantial scientific challenges. One key problem is limiting the production of harmful ice crystals during the freezing process. While cryoprotectants provide a solution, gaining maximum effectiveness, particularly in complex organs such as the brain, is a continuous effort.

Nanotechnology, or the manipulation of matter at the atomic and molecular levels, offers a promising way to overcome these constraints. Its prospective contributions center on improving cryoprotectant distribution and effectiveness, which could result in a dramatic leap in cryonics.

Microscopic robots, known as nanobots, show great potential for navigating the complicated human body with amazing precision. Unlike standard cryoprotectants with broad effects, these nanobots may provide an innovative approach:

In contrast to previous approaches, nanobots might be programmed to actively seek and reach specific cell types, particularly those most vulnerable to cryopreservation damage, such as delicate neurons. This focused strategy has the potential to greatly improve cryoprotectant efficacy when compared to existing broad-spectrum approaches, ensuring that even the most important structures are protected.

Nanobots' minuscule size could enable deeper cryoprotectant penetration into cells. This ensures that even the most delicate cellular structures are adequately protected, reducing any damage during the freezing process.

The potential significance of nanotechnology in cryonics goes beyond targeted cryoprotectant delivery.

Nanobots may continuously monitor numerous parameters of the cryopreserved subject, including temperature, cellular integrity, and other vital signs. This real-time data could be important for future revival efforts, allowing scientists to tailor their strategy and maximize their chances of success based on the individual's exact situation.

While it remains a distant possibility, advances in nanomedicine may pave the road for nanobots to play a role in the **potential healing of cellular damage** incurred during cryopreservation. This could include targeted manipulation of cellular structures or the administration of repair agents, increasing the likelihood of a successful revival in the future should such technologies become available.

While the possibilities opened up by nanotechnology are enticing, it is critical to recognize its early stages. These applications will require much research and development before they can become a reality. Furthermore, the ethical implications of such technology necessitate thoughtful analysis and open discussion. Responsible development and cautious control of potential misuse, unforeseen consequences, and equitable access are critical to ensuring that this technology eventually serves the greater good.

The human desire to understand life and death has always been a motivating force in scientific research. Today, advances in biological sciences are pushing the limits of this pursuit, revealing glimpses of a future in which our understanding of these fundamental notions could be dramatically altered. Exploring these possible advances, as well as their implications for cryonics and our understanding of life and death, is critical for developing educated discourse about humanity's future and the meaning of existence.

1. **Advanced Regenerative Techniques**: The potential of regenerative medicine, which uses stem cells and tissue engineering, has important implications for cryonics. If effective revival occurs, the ability to rebuild tissues and organs damaged during the cryopreservation process will be vital for returning persons to health. For example, advances in stem cell therapy may allow for the regeneration of neurons and other important tissues that were potentially damaged during cryopreservation, paving the door for a more full and successful recovery.

The potential applications of regenerative medicine go far beyond cryonics. Imagine a world in which people with organ failure or crippling injuries could receive organ transplants generated from their own stem cells, removing the need for organ donors and lowering the danger of rejection. This technology could also provide answers for age-related disorders such as macular degeneration, as well as neurological diseases such as Alzheimer's and Parkinson's, thereby enhancing the whole population's quality of life and life expectancy.

2. **Decoding and Editing the Human Genome**: The human genome, the blueprint for life, has the potential to reveal the mysteries of aging, disease, and perhaps death. With continued advances in genomics and gene editing technologies such as CRISPR, scientists may be able to comprehend and change the human genome with unparalleled accuracy. This opens up exciting possibilities for cryonics. Future therapies could potentially address genetic factors that contribute to cellular damage during cryopreservation, as well as age-related disorders, increasing the odds of successful revival and living a longer, healthier life afterward. However, the ethical implications of such technology necessitate

thorough scientific inquiry, rigorous ethical discourse, and responsible development to assure its safe and equitable usage.

The capacity to edit genes provides the door to a future free of inherited diseases, with the potential to fix genetic mutations that contribute to aging and age-related ailments. However, the ethical implications of this technology are enormous. Questions are raised regarding potential unintended consequences, the risk of creating designer kids with specified qualities, and guaranteeing equal access to this potentially life-changing technology. Open dialogue and responsible development will be critical in navigating this complicated landscape and ensuring that gene editing benefits human health and well-being.

3. **Biomimicry for Enhanced Preservation**: Biomimicry, or the replication of natural processes and structures for technological innovation, can provide useful insights into cryonics. Studying species that thrive in harsh settings, such as extremophiles, may lead to the creation of novel cryoprotectants or enhanced preservation strategies based on nature's resilience. Furthermore, biomimicry may inspire the development of improved monitoring systems

or even self-repair processes within cryopreserved individuals, drawing on the innate regeneration capacities of live organisms.

Biomimicry has the potential to transform several fields outside cryonics. By examining nature's answers to challenging challenges, scientists have created self-cleaning surfaces inspired by lotus leaves, energy-efficient buildings modeled after termite mounds, and even medicinal adhesives inspired by the strength of mussels. As we continue to investigate and learn from the natural world, the opportunities for creativity and issue solving expand exponentially.

4. **Investigating the Frontiers of AI**: The search to comprehend consciousness and its relationship to the brain is a complex and ongoing process. As artificial intelligence (AI) advances, the prospect of producing artificial consciousness or transferring consciousness from a biological brain to a non-biological substrate may emerge. While very hypothetical at the moment, this situation raises interesting considerations about the potential consequences for cryonics. Could consciousness be stored or even transferred in the future, thereby changing our view of life and death, as

well as the definition of personality itself?

The ability of AI to duplicate or even outperform human intelligence raises an intriguing philosophical issue. Questions are raised concerning the nature of consciousness, the notion of sensibility, and AI's potential impact on our society and our perception of what it means to be human. While the potential benefits of AI are apparent, it will be critical to ensure its responsible development and use for the betterment of humanity.

Chapter 4

Cryopreservation Process

Cryopreservation, the practice of preserving individuals at ultra-low temperatures with the hope of future revival, presents a captivating prospect within the field of biopreservation. While currently in its nascent stages, the procedure entails a meticulously orchestrated sequence of steps, each contributing significantly to this unique endeavor with uncertain future implications.

1. **Pre-Cryopreservation Preparations:** The initial stage commences with a comprehensive pre-cryopreservation evaluation. Conducted by a team of medical professionals, this evaluation ensures the individual's suitability for the procedure. It involves a thorough review of medical history, current health status, and identification of any existing medical conditions that might require specific considerations during cryopreservation. Open and

honest communication is paramount at this stage, as informed consent is obtained. This consent acknowledges the experimental nature of the procedure, its limitations, and the uncertain potential for future revival.

2. **Ethical and legal considerations** also come into play, requiring careful navigation of a complex landscape. Cryopreservation raises numerous questions about legal rights, future implications, and societal impact. Ensuring adherence to existing legal frameworks and fostering open dialogue is crucial. This process involves collaboration with legal professionals, ethicists, and other relevant stakeholders to address these complexities in a responsible and comprehensive manner.

3. **Pre-Freezing Procedures:** Following successful completion of the

preparatory phase, the pre-freezing stage commences. This delicate phase involves replacing the individual's blood with a **cryoprotectant solution**, a specialized chemical formulation acting as a cellular "antifreeze." Imagine these cryoprotectants as microscopic shields, safeguarding the delicate structures within cells from the harshness of extreme cold. The process, known as **perfusion**, is akin to a blood transfusion, carefully replacing blood with the cryoprotectant solution throughout the body.

Following perfusion, the body temperature is gradually lowered in a process called **controlled-rate freezing**. This slow and controlled descent mimics the natural process of entering hibernation, allowing cells to adjust and minimize stress on their intricate structures. Imagine this process as gently guiding the body into a state of deep slumber, ensuring minimal disruption to the cellular machinery.

3. Cryopreservation: The Core of the Procedure

With the body sufficiently cooled, the core of the cryopreservation process unfolds. Two primary methods exist, each with its own approach to achieving cellular preservation:

- **Vitrification:** This method, currently preferred due to its potential for minimizing cellular damage, aims to solidify the cryoprotectant solution and the biological material into a glass-like state, essentially bypassing the formation of ice crystals altogether. Imagine transforming the entire body, not into ice, but into a glassy suspension, effectively halting all biological activity and preserving the cellular structures in a near-life state.

- **Freezing:** The traditional method involves further cooling the individual to extremely low temperatures, typically around -196°C (-321°F), using liquid nitrogen. While simpler, this method carries a higher risk of ice crystal formation and potential cellular damage.

4. Storage: A State of Suspended Animation

Once cryopreserved, the individual is transferred to a specialized container filled with liquid nitrogen. This specialized chamber acts as a **time capsule**, maintaining the ultra-low temperature necessary for long-term preservation. Here, the individual rests in a state of suspended animation, awaiting potential advancements in technology and the chance of a future revival. Imagine the individual slumbering peacefully within this specialized chamber, their biological processes effectively paused, waiting the potential dawn of a future where revival becomes a reality.

It is crucial to acknowledge that revival from cryopreservation remains highly speculative at this stage. Significant scientific and technological advancements are necessary before such a feat becomes a reality. However, hypothetically, if and when revival technology becomes available, the individual would be gradually rewarmed to a physiological temperature. Depending on the condition of the individual and the advancements in medical technology at that time, various medical interventions might be necessary to support the revival process and restore complete functionality.

Challenges and Innovations

Cryopreservation, the practice of preserving individuals at ultra-low temperatures with the potential for future revival, presents a captivating avenue in the quest to expand the boundaries of human life. However, the intricate process faces significant challenges that hinder its potential for successful application. Addressing these challenges and fostering continual innovation are crucial for the advancement of this field.

1. The Ice Crystal Enigma: A Persistent Hurdle

One of the most formidable challenges in cryopreservation lies in the formation of ice crystals during the freezing process. These microscopic ice shards can inflict significant damage on delicate cellular structures, compromising the integrity of preserved tissues. This phenomenon, akin to the destructive effects of frost on delicate flora, poses a significant barrier to successful revival efforts.

Emerging Solutions: Orchestrating a Flawless Freeze

Scientists are actively exploring innovative solutions to combat this challenge:

- **Enhanced Cryoprotectants:** Ongoing research focuses on developing **more efficacious cryoprotectants**, specialized solutions that act as cellular antifreeze. These advancements aim to optimize the protective capabilities of these solutions, minimizing ice crystal formation and safeguarding the intricate structures within cells. Imagine these cryoprotectants as microscopic shields, evolving to offer even greater protection against the harsh realities of ultra-low temperatures.

- **Refined Vitrification Techniques:** As previously discussed, **vitrification**, the process of transforming the cryoprotectant solution and biological material into a glass-like state, offers a potential solution. Unlike traditional freezing, vitrification bypasses ice crystal formation altogether, potentially minimizing cellular damage. Ongoing research endeavors to refine these techniques for improved efficacy and broader applicability, striving to make vitrification a more robust and reliable method for cryopreservation.

2. Preserving the Cellular Symphony: A Delicate Balance

Another critical challenge involves preserving the **cellular integrity** during cryopreservation. The complex interplay between various cellular components is essential for life, and any disruption to this delicate balance could significantly impact the potential for revival. Imagine a sophisticated orchestra where every instrument plays a crucial role in creating a harmonious symphony. Cryopreservation aims to preserve this intricate symphony, ensuring that all cellular components remain intact and functional after the freezing process.

Innovation Takes the Stage: Protecting the Cellular Orchestra

To address this challenge, researchers are exploring various promising avenues:

- **Nanotechnology Applications:** The field of nanotechnology, dealing with the manipulation of matter at the atomic and molecular scale, offers intriguing possibilities. **Nanobots**, microscopic robots, could potentially be used for targeted delivery of cryoprotectants, ensuring optimal protection for vulnerable cellular structures. Additionally, nanobots might play a role in future repair strategies, potentially aiding in the restoration of any damage sustained during cryopreservation. Imagine these nanobots as miniature repair crews, meticulously working within cells to address any potential damage caused by the freezing process.

- **Advanced Monitoring Systems:** The development of **sophisticated monitoring systems** could become crucial for ensuring the optimal preservation of cryopreserved individuals. These systems could continuously monitor various parameters like temperature, cellular integrity, and other vital signs, allowing scientists to fine-tune storage conditions and potentially even guide future revival efforts.

Imagine these monitoring systems as watchful guardians, keeping a constant vigil over the cryopreserved individual, providing valuable data to inform future decisions and potentially guide the revival process.

Legal and Regulatory Landscape

Presently, there are **no specific laws or regulations** governing cryonics at the international level. This creates a somewhat **uncertain legal landscape**, with various jurisdictions approaching the issue differently. Some countries, like the United Kingdom, have no specific regulations, while others, like France, explicitly prohibit cryopreservation of people.

Existing Legal Considerations:

- **Consent:** Obtaining informed consent from the individual or their legal representative is paramount before cryopreservation. This consent acknowledges the experimental nature of the procedure and the uncertain potential for future revival.

- **Property Rights:** The legal ownership and disposition of cryopreserved individuals remains a complex question. Determining who holds the legal rights and responsibility for the individual and their eventual disposition requires careful consideration and potential legal clarification.

- **Ethical Concerns:** Cryonics raises various ethical concerns, such as the potential exploitation of vulnerable individuals, the allocation of scarce resources, and the broader societal implications of potentially extending life indefinitely. Open dialogue and responsible development are crucial to navigate these complexities.

Future Considerations: Charting a Course through Uncharted Waters

As cryonics continues to evolve, the legal and regulatory landscape might need to adapt to address emerging challenges and opportunities:

- **Standardized Regulations:** The development of **standardized international regulations** could provide clarity and consistency in the legal framework surrounding cryonics. This could involve establishing guidelines for informed consent procedures, property rights, and ethical considerations.

- **Public Discourse:** Fostering **open public discourse** and engaging with diverse stakeholders, including scientists, ethicists, legal experts, and the public, is crucial for responsible development and ensuring that cryonics aligns with societal values.

- **Global Collaboration:** Given the international nature of cryonics, fostering **global collaboration** among governments, scientific bodies, and ethical experts will be essential for developing a comprehensive and responsible regulatory framework.

Chapter 5

Potential Benefits for Maintaining and Enhancing Well-being

- **Motivation for Adopting Healthier Lifestyles:** The theoretical possibility of future revival, although uncertain, could incentivize individuals to **prioritize preventative healthcare practices**. This might translate to increased engagement in behaviors like regular exercise, balanced nutrition, and early detection of health issues. Focusing on physical and mental well-being could potentially improve overall health and **potentially enhance the chances of successful revival in the future**, should the technology become a reality. Research suggests that healthy lifestyle choices not only reduce the risk of chronic diseases but also improve cognitive function and mental well-being, contributing to a more fulfilling life in the present.

- **Mitigation of Existential Anxiety:** For some individuals, the existential anxiety associated with mortality can be a significant source of psychological distress. Cryonics, by offering a hypothetical chance at continued existence, could potentially offer **psychological relief** and a sense of agency in the face of mortality. This could lead to **improved mental well-being** and a reduced fear of death, allowing individuals to focus their energy on living a meaningful and enriching life in the present. Studies have shown that confronting mortality in a healthy manner can lead to personal growth, increased appreciation for life, and a stronger sense of purpose.

Potential Risks and Considerations Requiring Careful Navigation

- **False Hope and the Burden of Uncertainty:** The uncertain nature of future revival success and the extended timescales involved could lead to the creation of **unrealistic expectations and emotional distress**. Individuals might experience anxiety, disillusionment, or even depression if their hopes for revival are not met. This highlights the importance of **transparent communication** regarding the limitations and uncertainties surrounding cryonics. Responsible engagement demands acknowledging the experimental nature of the procedure and avoiding false promises that could lead to psychological distress.

- **Financial and Societal Considerations:** The costs associated with cryopreservation can be substantial, potentially placing a **financial burden** on individuals and their families. Additionally, the social implications of cryonics, such as the possibility of extended lifespans and its impact on societal norms, require careful consideration and **ongoing dialogue**. Open discourse involving diverse stakeholders, including scientists, ethicists, legal experts, and the public, is crucial for navigating these complex considerations and ensuring that any potential advancements in cryonics benefit society in a just and equitable manner.

- **Obsessive Focus on the Future at the Expense of the Present:** An excessive focus on the prospect of future revival could lead to **neglecting present-day well-being**. Individuals might prioritize potential future life over fully appreciating and engaging in the present moment. This underscores the importance of integrating cryonics with a **holistic approach to well-being** that emphasizes living a fulfilling life in the present. Cultivating gratitude for the present moment, building meaningful relationships, and pursuing personal growth remain essential for a flourishing life, regardless of any potential future prospects offered by cryonics.

A Balanced Approach Anchored in Transparency and Responsibility

Integrating cryonics with wellness demands a **nuanced and responsible approach**. While the potential benefits for physical and mental health should not be ignored, it is crucial to acknowledge and address the associated risks. This necessitates:

- **Open and Honest Communication:** Transparency about the limitations, uncertainties, and potential ethical complexities surrounding cryonics is paramount. Avoiding false promises and fostering realistic expectations are essential for responsible engagement.

- **Prioritizing Present-Day Well-being:** Cryonics should not detract from the importance of living a healthy and fulfilling life in the present. Maintaining physical and mental well-being throughout life remains the cornerstone of a holistic approach to health and happiness.

- **Ethical Considerations and Social Discourse:** Open dialogue involving diverse stakeholders, including scientists, ethicists, legal experts, and the public, is crucial for addressing the complex ethical considerations and societal implications of cryonics.

Ethical Implications of a Healthier Immortality

The potential to extend human lifespan and potentially achieve a state of "healthier immortality" through advancements like cryonics and regenerative medicine presents a captivating, yet ethically intricate, scenario. While the allure of a future free from the constraints of aging is undeniable, the path to achieving it necessitates careful consideration of the profound ethical dilemmas it presents.

Resource Allocation and Equity Concerns

- **Unequal Access and Limited Resources:** The potential for extended lifespan raises significant concerns about **equitable access to these life-extending technologies**. With limited medical resources and potentially exorbitant costs associated with cryonics and related advancements, ensuring equitable access for all individuals becomes a paramount challenge. We must critically examine how these advancements might exacerbate existing inequalities and create a scenario where only the privileged few can afford to extend their lives. Furthermore, intergenerational equity demands consideration. Will there be sufficient resources to support future generations if individuals continue to live and consume resources for extended periods? Striking a balance between ensuring access for all and safeguarding the well-being of future generations will be crucial.

Redefining Life, Death, and the Meaning of Existence

- **Existential Questions and the Redefinition of the Human Experience:** The potential to extend life indefinitely raises profound philosophical questions about the **meaning of life and death**. What happens to our values, motivations, and societal structures when the concept of mortality is fundamentally altered? Will a longer lifespan lead to a more meaningful existence, or could it foster feelings of ennui, purposelessness, and existential angst? Additionally, the possibility of significantly extended lifespan could lead to a **redefinition of the human experience**. What will it mean to be human when aging is no longer the inevitable end? How will our relationships, families, and societies adapt to this new reality? Open dialogue and critical reflection are essential for navigating these complex questions and ensuring that any advancements in lifespan extension serve the greater good.

The Need for Ethical Frameworks and Oversight

- **Developing Robust Ethical Frameworks:** As we delve deeper into this uncharted territory, robust **ethical frameworks** are crucial to guide responsible development and application of these technologies. These frameworks should address critical issues like informed consent, resource allocation, and potential risks associated with these advancements. Open and transparent public discourse involving diverse stakeholders, including scientists, ethicists, legal experts, and the public, is essential. This dialogue will allow us to collectively explore the potential benefits and risks, navigate the ethical complexities, and ensure that any advancement in lifespan extension are pursued in a manner that aligns with societal values and promotes the well-being of all individuals.

Chapter 6

Public Perception

While the concept of cryonics has garnered media attention in recent years, public awareness remains unevenly distributed. Surveys suggest that a significant portion of the population has either not encountered the concept or possesses only a fleeting familiarity. This lack of widespread knowledge can be attributed, in part, to the relatively nascent stage of cryonics research and development. Unlike established medical procedures, cryonics has yet to become a fully integrated aspect of public discourse.

Understanding Varied and Nuanced

Among those who are aware of cryonics, the level of understanding can vary considerably. Some individuals might possess a more nuanced comprehension of the scientific principles and limitations involved, acknowledging the complexities and uncertainties inherent in the technology. Conversely, others might hold incomplete or inaccurate information, shaped by sensationalized media portrayals or limited exposure to reliable sources. This spectrum of understanding underscores the need for comprehensive and accessible information that caters to diverse levels of prior knowledge and fosters critical thinking skills.

Navigating the Maze of Biases: Media, Science, and Ethics

Several factors contribute to the formation of public perception and can introduce potential biases:

- **Media Portrayals**: Media representations, often sensationalized or lacking scientific accuracy, can significantly shape how the public perceives cryonics. These portrayals, while captivating, can create unrealistic expectations or fuel negative stereotypes, hindering informed understanding.

- **Scientific Uncertainty**: The ongoing scientific and technological development surrounding cryonics inherently leaves room for uncertainty, which can lead to skepticism and hesitation among the public. This underscores the importance of transparent communication about the current state of the technology and the ongoing research efforts aimed at addressing existing limitations.

- **Ethical Concerns**: Cryonics raises various ethical questions, such as informed consent, resource allocation, and the potential implications for society. This ethical complexity can generate public debate and mixed perceptions, highlighting the need for open dialogue and collaboration among various stakeholders to navigate these considerations responsibly.

Chapter 7

Conclusion

As we stand at the precipice of a future potentially reshaped by cryonics, the words of Arthur C. Clarke ring true: "The only way to discover the limits of the possible is to venture a little way past them into the impossible." While the success of cryonics remains a question mark etched in the unknown, the pursuit of this knowledge necessitates a collective effort guided by scientific rigor, ethical considerations, and unwavering curiosity. By fostering open dialogue, embracing interdisciplinary collaboration, and approaching this field with a spirit of responsible exploration, we can ensure that the potential of cryonics is investigated and, if possible, realized for the benefit of all humanity.

Let us not shy away from the complexities or the uncertainties. Instead, let them serve as catalysts for critical thinking, open discourse, and a commitment to shaping a future grounded in scientific inquiry, ethical responsibility, and the enduring human spirit of exploration. This journey into the uncharted territory of cryonics demands not just scientific advancements, but also a collective responsibility to confront the ethical dilemmas head-on. We must engage in open and inclusive dialogue, encompassing diverse stakeholders from scientists and ethicists to legal experts and the public at large. Through this collaborative approach, we can ensure that the potential benefits of cryonics are pursued in a manner that aligns with our societal values and promotes the well-being of all individuals.

As we delve deeper into this captivating field, let us remember the profound words of Carl Sagan: "Somewhere, something incredible is waiting to be known." While the path ahead may be fraught with challenges, the potential to unlock the mysteries surrounding life and death compels us to continue this exploration with intellectual rigor, unwavering curiosity, and a commitment to responsible progress. The future of cryonics lies not solely in the hands of scientists and researchers, but in the collective responsibility and collective curiosity of humanity as a whole. Let us embrace the unknown with open minds, engage in meaningful discourse, and embark on this journey together, ensuring that every step we take is guided by the principles of scientific inquiry, ethical responsibility, and the enduring human spirit of exploration.